HIGH AND DRY

How Free Trade in Water Will Cripple Australian Agriculture

HIGH AND DRY

How Free Trade in Water Will Cripple Australian Agriculture

Patrick J Byrne
Neil Eagle
John O'Brien
Daryl McDonald

2006
Freedom Publishing

First Published October 2006
Reprinted December 2006
Freedom Publishing
582 Queensberry Street
North Melbourne. Vic 3051.
nw@newsweekly.com.au
www.newsweekly.com.au

Cataloguing-in-Publication Data:

HIGH AND DRY: How Free Trade in
Water Will Cripple Australian Agriculture
      Byrne, Patrick J; Eagle, Neil;
      O'Brien, John; McDonald, Daryl.

ISBN 0-9775699-2-6

Cover and book design by Eastern Advertising
Edited by Patrick J Byrne
Printed by Brougham Press, 33 Scoresby Rd, Bayswater,
Victoria. 3153.

Contents

Preface

This book has been written to give farmers a voice. Their farm peak bodies and their political representatives have failed them badly in the debate on water policy.

This book is written to awake policy makers and their advisors to the impending agricultural disaster if they continue to pursue open water trading markets.

There are alternatives to the National Water Initiative's water trading policies, but they demand attention to theoretical and practical details that have been ignored in the development of current water policy.

This book arose out of the concern of the authors, as well as many farmers and rural and regional community leaders, who have been dismayed at the path taken by governments under the National Water Initiative.

This book is adapted from a recent submission to the Productivity Commission inquiry on water trading and environmental water use.

About the Authors

Pat Byrne is Vice-President of the National Civic Council, and is a consultant to a number of farm industry groups.

John O'Brien has been a Gippsland mixed farmer for 45 years. He is the former head of the Victorian Farmers Federation water committee. He has served on water catchment authorities and rural water authorities, and has extensive experience in resource and catchment management.

Neil Eagle was a member of the Anomalies Committee for the Murray River Private Diverters, which developed the formula for Water Allocations from the Murray River. This became the template for all NSW river allocations. He has been involved in agriculture and water issues for over 50 years.

Neil is a citrus and beef producer near the Murray River at Barham, NSW. He is the former Chairman of Australian Citrus Growers Inc. (the citrus industry peak body) and remains on that board; Chairman of Mid-Murray Citrus Growers for 20 years; and since 1972, Chairman of Border Packers Pty Ltd., a grower-owned citrus packing and processing company.

Neil was chairman of the Murray-Lower Darling River Management Board for 11 years (1986-1997); and community representative on the Barmah-Millewa Forest Community Consultative Committee with the Murray Darling Basin Commission, that formulated the Barmah-Millewa Forest watering scheme.

Daryl McDonald is a mixed farmer near Swan Hill. He has been an irrigator and Murray River recreational water user and conservationist for 30 years. His family has a history that spans four generations, over 120 years, of irrigation farming adjacent to the Murray River.

Glossary

ABARE – Australian Bureau of Agricultural and Resource Economics

BCA – Business Council of Australia

COAG – Council of Australian Governments

FAO – Food and Agriculture Organization of the United Nations

GMW – Goulburn-Murray Water

MDB – Murray-Darling Basin

MDBC – Murray-Darling Basin Commission

MDBMC – Murray-Darling Basin Ministerial Council

NSW – New South Wales

NWI – National Water Initiative

NWC – National Water Commission

OECD – Organisation for Economic Cooperation and Development

PC – Productivity Commission

UN – United Nations

WSAA – Water Services Association of Australia

RECOMMENDATIONS

Recommendation 1. That the Federal Government undertakes an immediate overhaul of the National Water Initiative and the National Water Commission, putting in place advisors with expertise in irrigation water management and catchment area management so as to put order back into irrigation farming before current policies do major damage to Australian agriculture.

Recommendation 2. Governments must ensure that the water markets for agriculture, urban, industrial and environmental uses remain separate markets, with separate prices based on delivery costs. Governments must recognise that water is a mixed good, not a private good, with many public good characteristics. Because water is a mixed good, with different characteristics in different markets, only governments can decide *primary water allocations* among these markets. It cannot be allocated through a single, open trading market, selling water to the highest bidder. Particularly given the current chronic drought, governments must immediately quarantine irrigation water so that it cannot be permanently traded for urban, environmental or other uses, so as to ensure security of supply to the agricultural sector.

Recommendation 3. Governments should recognise that the idea of trading water from low value to high value agriculture is a fundamentally flawed concept. The final value of product to the economy depends on down stream processing, not merely farm gate price. What is a high value product today, can quickly become tomorrow's low value product, as is currently happening in the wine grape industry. Further, excessive emphasis on "high value" farm gate price products conflicts with other economic and social objectives, like providing the Australian people with low cost food, and helping Australian farmers achieve a competitive advantage over their heavily subsidised foreign competitors. Governments must recognise that it is better to shift agriculture to available water than to shift water to agriculture; that this is the common practice of farmers; and that farmers are in the best position to judge water use for "low-value" and "high-value" crops.

Recommendation 4. Governments must ban permanent water trade out of catchment areas. Trade of secure irrigation water should be restricted to trade among farmers within a catchment area.

Recommendation 5. The Federal Government must restrict managed investment schemes to timber plantations in high rainfall regions above 640 millimetres annually, and apply environmental guidelines to ensure that new plantations do not adversely reduce water flows in catchments, as well as the supply of water to other forms of agricultural and other water users.

Recommendation 6. It is imperative that governments implement the findings of the House of Representatives Standing Committee on Agriculture, Fisheries and Forestry *Inquiry into future water supplies for Australia's rural industries and communities – Interim Report:*

Recommendation 1: In light of the Committee's severe reservations about the science, the Committee recommends that the Australian Government urge the Murray-Darling Basin Ministerial Council to postpone plans to commit an additional 500 gigalitres in increased river flows to the River Murray until:

• a comprehensive program of data collection and monitoring by independent scientists is completed;

• non-flow alternatives for environmental management are considered and reported upon more thoroughly; and

• a full and comprehensive audit focused specifically on the Murray-Darling Basin's water resources, including all new data, is conducted.

Recommendation 2: The Committee recommends that the Australian Government ask the Murray- Darling Basin Ministerial Council to allocate sufficient funds out of the $500 million allocated to the River Murray by COAG to the abovementioned tasks, prior to proceeding with the proposal to obtain increased river flows.

Recommendation 7. Where thorough, community agreed science shows that there is a need for environmental flows in rivers, then governments must create a separate environmental water market. Water and land set aside for environmental purposes are public goods. Water for these flows is to come from water savings from infrastructure improvements and from tapping new water supplies, not from purchasing irrigation water from farmers. The cost of environmental flows should be borne by the whole population collectively, as they are the environmental custodians of our river environments.

Recommendation 8. Governments must recognise that there is no single ideal form of on-farm irrigation delivery system and that farmers should be provided financial incentives to invest in the most efficient forms of irrigation for their particular farm products.

Recommendation 9. Rather than using price penalties to encourage water savings on farms, governments should offer positive financial incentives to farmers for on farm enhanced water savings and intensification of agricultural land use, including where this results in parts of farmland being transferred from marginally profitable farming to being used for environmental purposes. Positive incentives should be offered, just as the Federal Government provides financial incentives to the states to implement competition policies under National Competition Policy.

Recommendation 10. The Federal government should provide financial incentives for the states to undertake a full audit of irrigation entitlements, particularly ground water entitlements.

Recommendation 11. The Federal Government must hold the states accountable for National Water Initiative payments, ensuring that these payments are used in water savings infrastructure and/or for the construction of new water storages, and not diverted into general revenue.

Recommendation 12. Governments should develop new water infrastructures in regions where there is ample water supply. Population migration should be promoted by grants and assistance for business transition arrangements so that skills in irrigated farming and irrigation support services are moved to areas where expansion of irrigation is possible. This would reduce pressures on regions with insufficient water supply.

Recommendation 13. The Federal Government needs to ensure that state and territory audits of available water include examination of the huge untapped resources in northern Australia (across north Queensland, the Northern Territory and the north of Western Australia), as well as substantial untapped water in NSW and Victoria. This audit should include identifying areas for the future expansion of Australian agriculture.

Recommendation 14. The Federal government must provide major new financial incentives for the states to build new water storages and delivery systems.

1

Overview

By the time they have finished reading this book, readers should come to see that:

- water trading has become a device for taking farmer's water and diverting it to towns and cities, forcing up the price of water;
- free trade in water is being rorted by managed investment schemes (MIS) which are buying up farmers' water rights as a tax dodge, further pushing up the price of water;
- more water is going to be taken from farmers for environmental flows down the Murray River, before the science on river health has been done. This will also force up water prices; and
- if the policies advocated continue to be implemented, it will devastate irrigation farming, push farmers off the land and push up the price of food for people in the cities.

National Water Initiative

This impending disaster is the result of National Water Initiative (NWI) policy which seeks to separate farmers' water entitlements from their property title for the purpose of open water trading between agricultural, urban, industrial and environmental use. This policy also proposes freeing up water trade to boost agricultural production, by moving water from "low value" to "high value" agriculture.

The NWI was started in the early 1990s, under the auspices of the Council of Australian Governments (COAG) and has been administered by the National Water Commission. It has developed to become the main focus of Australia's water policy, which has come to be both a rural and an urban issue because of looming water shortages.

There are four pillars to the NWI policy on irrigation water trade:

- an open, competitive water market can efficiently and economically allocate water among agricultural, urban, industrial and environmental users;
- unstated in the debate, but important to understanding the fatal flaws in the policy, is the assumption that water is a private good, not a public or mixed good;
- open water trade among irrigation areas will see water flow from low value to high value agriculture, boosting the value of Australia's agricultural output; and
- creating substantial environmental flows, particularly for the Murray River, are necessary to improve river health.

Contrary to the first pillar, research by the Food and Agricultural Organization (FAO) of the UN and World Bank recognises that the *primary allocation* of water resources is a political decision by governments to be determined by how a country wishes to strategically allocate its water among various water markets – rural, urban, industrial, environmental, etc. Both argue that economic efficiency methodologies (like cost-benefit analysis) have objectives that assist in the fine-tuning of water allocations, but only within a particular market. Both argue that economic efficiency methodology – under what the Productivity Commission calls "allocation efficiency of water use" – cannot determine the *primary allocation* of water resources.

Yet the Productivity Commission, a major player in the water debate, argues that the "allocation efficiency of water use" is concerned with allocating water among agricultural, urban, industrial and environmental uses,[1] to ensure "that the community gets the greatest return (broadly defined) from its scarce resources".[2] This is the broad policy position of the NWI, the Business Council of Australia and a number of free market think tanks.

Both the FAO and the World Bank research argues that *primary allocations* are to separate markets, each requiring different pricing arrangements. These involve decisions by governments.

Oils ain't oils

Attempts to devise open water markets are encountering frequent problems. This is reflected in a number of Productivity Commission papers.[3] Trying to integrate rural, urban, industrial and environmental markets, then expecting economic forces to allocate water among these markets and set prices, encounters major market failures. This arises from conceptual failures to recognise that water, a mixed good, has different characteristics in different markets, and cannot be traded like private goods in the free market.

Just as "oils ain't oils" in the old TV advert , "water ain't water".

Conveniently, in the period when Australia's population and agriculture were both expanding substantially, there was ample water for rural, urban and industrial use, particularly during Australia's relatively wet period from around 1945 to the 1980s. Urban water was provided at the cost of purification and reticulation; rural water at the cost of delivery.

Significantly, both rural and urban water markets were able to operate separately, without competition. Storages were constructed primarily for irrigation use or for urban use, with limited overlap. Now, governments are seeking another market for water, environmental flows down rivers.

In the Snowy Mountains, the system of dams, weirs, locks and channels was constructed to provide an irrigation system through the Murray-Darling Basin. Across Australia, just 2 per cent of agriculture generates 80 per cent of the profits, and most of this comes from irrigation agriculture in the Murray-Darling Basin[4]. Significant irrigation systems were built in other states. The objective was to construct a large agricultural industry, to make Australia self-reliant in food, a major food and fibre exporter and a leader in many areas of agricultural technology. Adding value to this product, the largest section of Australian manufacturing industry involves the processing of food and fibre from Australian farms.

To achieve this objective, irrigation water has been provided at delivery cost and not in competition with urban water. Until now, irrigation water has effectively been quarantined. Low cost water has resulted in:

- low cost, quality food to the domestic market, with virtually no subsidies;
- a competitive edge for Australian farmers on the world market, while the rest of our developed world competitors receive substantial direct subsidies; and
- a significant proportion of Australia's annual export revenue.

Over the past 80 years, Australia built large and separate water storages for agriculture and cities. It also developed an accompanying complex, yet integrated system of physical infrastructure, legal rights and localised water use rules. They have been geared to regional soil and environmental issues, detailed local farming and environmental knowledge, and social infrastructure. Together, these provide a finely tuned system in each region of the Murray-Darling Basin, attuned not just to the economic needs of the nation, but the demands of local conditions, with varying soils, vegetation, habitat, salinity and other environmental conditions. Indicative of the complexity of the irrigation water market, there are 24 different types of distinct water rights, with different characteristics, security levels and resources to back appropriate delivery entitlements.

Again, even with irrigation water, just as "oils ain't oils", so too "water ain't water".

The result has been a valued agricultural contribution to the economy, and a good record of environmental stewardship.

To take a legal analogy, this intricate irrigation system of the Murray Darling Basin is comparable to the system of common law developed over hundreds of years. It has been moulded and adapted to suit the climate and landscape of each region and, in turn, the types of agriculture each region can sustain.

Why the fatal policy shifts?

What has changed to cause governments to unbundle 80 years of successful, scientific-based agricultural rules, property and water rights, neglecting enormous, accumulated institutional knowledge, to now allow the trade of water from farms to cities, and among irrigation regions?

First, since the 1980s, Australia has experienced a relatively dry period, possibly the prelude to a prolonged drier spell comparable to the period 1920 to 1945. As the demand for water has risen with urban growth and the precipitation rate has fallen, governments have failed to examine the feasibility of constructing new reservoirs. Indeed, some states have rejected new water storages as part of their official water policy. Instead, they are pursuing the cheap option – trading water from agriculture to the cities.

Notably, the NWI makes no reference to the possibility of expanding water storages; nor has the COAG any policy of financial incentives under National Competition Policy for the states to build new storages. Instead, they have focused financial incentives on deregulating water markets to allow water to be taken from farmers to overcome shortages in urban areas.

Indeed, the Productivity Commission's draft *Review*

of National Competition Policy Reforms (2004) states that a key opportunity arises from deregulation of the water market, namely that of "better integrating the rural and urban water reform agendas, including through facilitating water trading between rural and urban areas". This is regarded as the easiest means of solving water shortages because irrigation water was regarded as "probably the cheapest source of new water for several cities ..."[5]

What policy makers also fail to understand is that the price curve for water is inelastic. Hence in tight markets, a small increase in demand, or a small decrease in supply, can cause the price of water to rise dramatically. Irrigation farmers cannot sustain a substantial permanent rise in the price of water.

Invariably over time, as Australia's population expands and farmers retire or face hard times, their irrigation water will be sold off to the highest bidders for urban, environmental and industrial use. In drought times, some will sell off their permanent water right in a desperate effort to survive, as is currently happening in some regions. Others will be forced out as the price of water is raised to the point where they can no longer survive. Others will be forced out no matter how efficient they are, because their assets will be stranded along irrigation channels where it is no longer economic to supply water after other farmers have sold their water rights out of the district. A critical mass of farmers will be lost in one irrigation area after another. This scenario will see many irrigation regions facing extinction.

This is the medium-to-long-term logic of integrating agricultural, industrial, environmental and urban water markets and allowing water to be traded to the highest bidder at the highest price. The policies being proposed are short-sighted and threaten the viability of large sections of irrigation agriculture, with dire economic consequences not just for farmers, but for the economy. Agriculture and its dependent (input and output) industries make up 12.2 per cent of the Australian economy, and open water trading puts its most productive regions are under threat.

Deregulation of irrigation agriculture to supply water to the cities is evidence of the failure of state water policies to provide water from Australia's ample untapped water resources. It is bad agricultural policy and bad economic policy.

Second, it has been argued that the tying of water entitlements to property has inhibited the ability of water to be traded from "low value" to "high value" agriculture. Economic modelling claims to show big increases in the value of agriculture following water trade. However this modelling appears to be seriously flawed.

Most permanent water trade from one catchment to another has been to managed investment schemes for timber, wine grapes, olives and other products. Here, the shifting of water is not based on market forces for the product on these farms, but on tax breaks for Collins Street investors. Because these investments are not a response to market forces, over investment can rapidly turn today's "high value" farm products into tomorrow's "low value" products. Moreover, water purchases on behalf of these schemes are distorting the market and creating water barons capable of manipulating future water prices.

Policy makers have assumed that high value products are restricted to areas of the Murray-Darling Basin that are short of water. This is false. Most high value products can be produced almost anywhere in the basin. As will be argued below, it is better to shift the high value agriculture to the water, rather than the water to high value agriculture. In fact, this has been the practice of farmers.

Policy makers seem to believe that farmers are slow to invest in new technologies and crops, and that water trade is needed to encourage new investment. This is also false. In fact, as Productivity Commission research has shown, agriculture is the third highest productivity sector in Australia, after the communications and wholesale sectors. Farmers have been willing to adopt new irrigation technologies and improve productivity, although the recent long dry spell has slowed investment.

From the farmers' perspective, the urgency is to develop markets for their products, not water-trading. They can grow almost any product almost anywhere in the Murray-Darling Basin. The problem is finding markets for what they can produce.

For these reasons, unbundling of water entitlement from land title to promote trade in irrigation water is entirely unnecessary. It is like cracking a peanut with a sledgehammer.

Third, it is being argued that water deregulation is necessary in order to buy water for environmental flows, to improve the health of rivers like the Murray. Yet, the science of Murray River health is so poor that demanding water deregulation to provide environmental flows is a very premature step. Members from all political parties are represented on the House of Representatives Standing Committee on Agriculture, Fisheries and Forestry, which issued an interim report on The Living Murray environmental flows proposal in 2004. Its members voted 11:1 to reject proposals for 500 gigalitres annual environmental flow down the river. It said that the science on environmental flows, just one of 22 issues in river health, was so poor that plans for the 500 gigalitre environmental flow should be postponed until a comprehensive program of data collection and independent monitoring was completed; and until other alternatives to river management were examined.

Any decision on the use of environmental flows, as proposed in The Living Murray, are premature. Deregulation of irrigation water to allow purchases for environmental flows, when full scientific considerations may show such flows as unnecessary, is bad policy. Eventually, if good scientific evidence is found for environmental flows, then a separate environmental water market needs to be created and the water should be found from savings in the basin, not by taking or buying water from farmers.

Like abolishing common law

To continue the common law analogy, the unbundling of

water entitlements and deregulation of water trade are analogous to the federal and state governments agreeing to abolish, via simultaneous acts of parliament, the entire common law system and replace it with a bill of rights, arguing that the bill of rights will guarantee a fair and just legal system in which everybody's rights will be respected, while making the administration of the legal system easier and less complex. Perhaps the new system will work better than the old common law. However, it is impossible to imagine such a fundamental change taking place without a major public debate over a long period of years to weigh up the pros and cons.

The changes proposed to irrigation agriculture are of comparable significance and economic magnitude. But, whereas Australians have some understanding of common law, very few have an understanding of the complexities of irrigation agriculture, or realise what is at stake if the proposed NWI policies proceed.

So far the debate has been fatally flawed, yet policy is advancing rapidly.

If the key policies being advocated are implemented, then the genie will be out of the bottle and it will be extremely difficult and very costly to repair the resulting damage.

Proposed sweeping plans to deregulate the water market and allow water trade between rural and urban markets, to trade so-called "high value" agriculture and to create environmental flows are all fundamentally flawed concepts. Consequently, the NWI policies:

- address only the needs of urban and industrial water users at the expense of irrigation farming, while ignoring the feasibility of building new reservoirs;

- are unnecessary in order to foster high value agriculture;

- fail to demonstrate from sound, community agreed science that only environmental flows produce pristine rivers, when river flows are just one of 22 issues affecting river health; and

- threaten the future of agriculture and its dependent industries, which make up 12.2 per cent of the Australian economy.

The process has only just begun. It is not too late for governments to change policy direction and stop undermining irrigation agriculture.

Recommendation: That the Federal Government undertakes an immediate overhaul of the National Water Initiative and the National Water Commission, putting in place advisors with expertise in irrigation water management and catchment area management so as to put order back into irrigation farming before current policies do major damage to Australian agriculture.

Endnotes

1 *Rural Water Use and the Environment: The Role of Market Mechanisms: Discussion Draft*, Productivity Commission, 2006 p. 212.

2 Ibid., p. 207

3 *Rural Water use and the Environment: The Role of Market Mechanisms: Discussion Draft*, Productivity Commission, 2006. *Stranded Irrigation Assets: Staff Working Paper,* Productivity Commission, June 28, 2006.

4 *The Living Murray,* Murray-Darling Basin Ministerial Council, July 2002. p. 16.

5 *Review of National Competition Policy Reforms: Discussion Draft,* Productivity Commission, October 2004, p. 209, 206.

2

Water Rich, Policy Poor

Contrary to popular opinion, Australians are among the most water rich people in the world. Although the water is not where the population has settled, Australia has options that largely have been ignored in the current debate. The failure of governments over the past two decades to plan for population expansion and for Australia's periodic dry periods has led to shortages and a knee jerk reaction – open water trade to allow the sale of irrigation water from farms to cities, industries and the environment. Open, free trade in water is the biggest immediate threat to Australia's irrigation agriculture, which makes up one-third of Australia's gross agricultural output.[1]

Instead of asking, "how much water is needed" and "where is it needed" – and planning ahead for irrigation, urban, industry and environmental needs – there has been gross political neglect. Shortages have been blamed on drought. "Australia is a dry continent, so we have to learn to economise" has become a mantra and a lame excuse for the failure of politicians and planners.

Based on the popular narrow view that Australia is a "dry" continent, environmentalists, bureaucrats and politicians have pursued the policy line of least resistance on the issue of water shortages. They have focused on the lowest cost, short term solution – letting urban areas buy water from farmers. Governments have bowed to the environmental lobby and opposed the building of new reservoirs.

Not short of water

Australia's water resources should be put in perspective. *The Macquarie Atlas* provides an overview, saying,

> "While Australia can rightly be described as the world's driest continent, it does not follow that the country is short of water. On a per capita basis, few countries in the world are so favourably endowed."

Australia is still to learn a lot about its true reserves.

> "Large areas are still not gauged and about half of the continent's estimated surface runoff comes from ungauged catchments. Given the large dependence on estimates, the figures are subject to a high degree of error, possibly by as much as 30 per cent in northern Australia ..."

While Australia has the lowest precipitation and run off of all continents, a quarter of the continent across northern Australia "contributes over 80 per cent of total run-off."

The Atlas says that Australia has an annual run-off of 3,997 million megalitres. (One megalitre is the amount of water in one Olympic size swimming pool). Of this, the total divertible fresh water yield is estimated at about 100 million megalitres. Of this only about one-fifth, or 21.5 million megalitres, has been tapped.[2]

Similarly, only rough estimates can be made about Australia's groundwater reserves. Large areas of Australia are awaiting groundwater exploration.

The Macquarie Atlas estimates that of divertible fresh groundwater, only about 12.5 per cent, or 2.6 million megalitres of the available 30.3 million megalitres, is used annually. The major groundwater resources are found in northern Australia and the Murray-Darling Basin.[3]

When the Council of Australian Governments (COAG) met to advance the National Water Initiative (NWI) in 2004, Jennifer Marohasy, director of the Environmental Unit of the Institute of Public Affairs, argued that the governments' water policies should start with the knowledge that Australia is one of the most water-rich nations on earth. Marohasy said:

> "Delegates attending the COAG meeting would do well to approach the National Water Initiative with positive pragmatism ...

> "According to the World Resource Institute, we have 51,000 litres of available water per capita per day. This is one of the highest levels in the world, after Russia and Iceland, and well ahead of countries such as the United States (24,000) and the United Kingdom (only 3,000 litres per capita per day) ...

> "Many businesses are doing it tough, particularly in areas of continuing drought and where there is no developed water infrastructure. The bottom line, however, is that as a nation we are not really that thirsty."[4]

Much of Australia's agriculture is in the Murray-Darling Basin. It has an average annual flow of 22,700 gigalitres, about 6 per cent of Australia's annual fresh water run-off. (One gigalitre is 1,000 Olympic-size swimming pools). About 11,431 gigalitres is diverted, mostly for agriculture and some to supply Adelaide. The basin produces about 40 per cent of Australia's agricultural production.[5]

In contrast, much of Australia's fresh water run-off is above the Tropic of Capricorn. Yet, north Queensland (excluding the super-wet belt) produces only about one per cent of the nation's agriculture, while having far more water available than the Murray-Darling Basin.

Consider the magnitude of the available water in northern Australia's river systems, against that of the Murray Darling Basin:

- The Murray-Darling Basin (MDB), 22,700 gigalitres.
- Queensland's north-east coast: 91,500 gigalitres – four times that of the MDB.
- Gulf of Carpentaria: 130,500 gigalitres – 5.7 times the MDB.
- The Timor Sea rivers: 81,200 gigalitres – 3.5 times the MDB.[6]

Some of the northern rivers that make up these massive flows could be tapped, using diversions and some dams, without threatening their environmental flows.

One option would be to develop areas of northern Australia and, over time, shift more agriculture into these regions.

An alternative would be to take advantage of the natural features of the Australian landscape. Eastern Australia tilts inland towards Lake Eyre, towards the centre of the continent. Intermittently, tropical cyclones dump vast amounts of rain in north Queensland beginning a long flow of water down the back of the state, through the south-west channel country and eventually into Lake Eyre. This tilt of land means that water can be channelled south using gravity flow, partly onto the large "black plains" soils of western Queensland or further south to the top end of the Murray-Darling Basin.

While the great bulk of precipitation falls in northern Australia, there are other large untapped water supplies down the east coast. In NSW there are huge flows down the Clarence River system.

In Victoria, the Gellibrand River, off the Otway Ranges just south of Geelong, has a current annual flow of 296,000 megalitres. Furthermore, the Mitchell River has a current average flow of 500,000 megalitres,[7] while the Ovens and King rivers are lightly utilised.

Looking in the wrong direction

This leads to several observations. The claim that Australia is short of water is far from the true situation.

It would be impossible to drought proof the whole of Australia, however, long-term planning and investment should make it possible to drought proof our cities and to greatly reduce the impact of drought on irrigation agriculture. The claim that drought, or climate variations or climate change are the cause of current problems is an excuse for years of neglect and a lack of public investment in water resources.

The NWI has been operating for over a decade. The Federal government has provided considerable funding for various initiatives, and the states have been the major recipients of much of this funding. Aspects of Federal and State water planning under the NWI are both sensible and overdue, particularly in terms of water savings from infrastructure improvements by councils, households and industry.

Curiously however, under the NWI there have been no serious incentives put in place to comprehensively map state and national water resources. There have been no incentives put in place for the building of new reservoirs.

Instead, a major emphasis has been on water trade from rural to urban users. Underlying this policy is the belief that much of Australia's water shortage issues can be worked out by having the primary allocation of water between rural, urban, industrial and environmental use achieved through water trading. The attitude has been, "Let water be traded between these primary markets to the highest bidder. Let water be traded from low to high value agriculture. Let water be traded from farms to cities. Then governments won't have to undertake major new expenditures on new reservoirs and delivery systems."

Policy makers have wrongly assumed that water is a private sector good that should be traded to the highest bidder.

These assumptions about the nature of water and water markets underpin the NWI, and strongly influence research by the Productivity Commission, the Business Council of Australia and various economic think tanks.

Extraordinarily, there has been a deafening silence from many of Australia's leading farm organizations, few of which seem to have grasped the threat these policies now pose to the future of their members.

Recommendation: The Federal Government needs to ensure that state and territory audits of available water include examination of the huge untapped resources in northern Australia (across north Queensland, the Northern Territory and the north of Western Australia), as well as substantial untapped water in NSW and Victoria. This audit should include identifying areas for the future expansion of Australian agriculture.

Recommendation: The Federal government must provide major new financial incentives for the states to build new water storages and delivery systems.

Endnotes

1 "Irrigated farms generate one-quarter of Australia's ag output", *FarmOnline,* 27 September 2006, www.farmonline.com.au/news_daily.asp?ag_id=37586

2 *The Macquarie World Atlas,* The Macquarie Library Pty. Ltd., 1995, p. 158.

3 Ibid., p. 159.

4 Marohasy, Jennifer, *Australian Financial Review*, June 26, 2004.

5 *The Living Murray: A Discussion Paper on Restoring the Health of the River Murray,* Murray-Darling Basin Ministerial Council, July 2002, p.1.

6 Byrne, Patrick, "Not so dry continent: Australia has water options," *News Weekly,* September 11, 2004.

7 *Central Region Sustainable Water Strategy: Appendix 2,* Draft Strategy, Victorian Department of Sustainability and the Environment, April 2006, p. 156.

3

Primary Water Allocation has to be done by Governments

The first pillar of the National Water Initiative (NWI) is to allow open, competitive markets in water trade to make the *primary allocation* of water among agricultural, urban, industrial and environmental markets. For example, the Productivity Commission (PC) argues that open water trading is needed to "allocate 'water to the highest use'[1] ... [This is to ensure] that the community gets the greatest return (broadly defined) from its scarce resources."[2]

Research by the both the Food and Agriculture Organization (FAO) of the UN and the World Bank strongly argues the contrary, that the *primary allocation* of water between markets can only be done by governments. It cannot be achieved by open trade in water.

An FAO research paper provides an extensive discussion on the valuation of water resources in agriculture, and warns against using market forces to price and allocate water. In its report, *Economic valuation of water resources in agriculture,*[3] it says:

> "Although water resources perform many functions and have important socio-economic values,

water is in many respects a classic non-marketed resource ...

"Economics is anthropocentric [human centred], and as such provides useful tools that can support decision-making. However, decisions concerning water allocations are guided not only by concerns of economic efficiency but also considerations of equity, environmental protection and social and political factors, to name but a few."

A report for the World Bank, *Water Allocation Mechanisms: Principles and Examples,*[4] also strongly argues that primary allocation of water resources requires decisions by governments. The report says:

"... three main points support the argument for public or government intervention in the development and allocation of water resources: it is difficult to treat water like most market goods, water is broadly perceived as a public good, and large-scale water development is generally too expensive for the private sector ...

"The state's role is particularly strong in intersectoral allocation, as the state is often the only institution that includes all users of water resources, and has jurisdiction over all sectors of water use ...

"Many countries adhere to some form of Public Trust Doctrine, that maintains that the state holds navigable waters and certain other water resources as an aspect of sovereignty. Because these are held as common heritage for the benefit of the people, the state cannot alienate such ownership of the basic resource and concomitant responsibility (Koehler 1995)[5]. This argument has been used in both the United States and India as a basis for environmental protection (Moench 1995)[6] – reasserting the state's role in allocating water between agricultural, industrial, municipal, and environmental sectors."

The FAO says that although the economic efficiency argument, on which the NWI strategy is based,

"is an important factor, there are additional economic issues that decision-makers need to consider. Two of these issues are the distribution of costs and benefits across society and their distribution across generations."[7]

All these issues require decisions by government. For example, the cost of reservoirs, irrigation channels, water treatment and city delivery systems should be spread across the generations that benefit from these infrastructure works. This can be done by governments, but cannot be done by the private sector. In recent times there has been a tendency to discount the cost of such infrastructure over a 20-year period, when the life of a dam may be 100 to 150 years, or more. Conveniently for some states that have a policy of "no new dams", pricing water to recover costs over 20 years means that new reservoirs are considered "too costly" to consumers.

Like the FAO report, the World Bank report says that:

"While economic efficiency is concerned with the amount of wealth that can be generated by a given resource base, equity deals with the distribution of the total wealth among the sectors and individuals of society. Many forms of water allocation schemes attempt to combine both efficiency and equity principles."[8]

Again, allocating water so as to keep equity and efficiency principles in balance requires decisions by governments. Such decisions cannot be done by markets.

The World Bank report goes on to argue the advantages and disadvantages of government involvement in the primary allocation of water to different markets. Despite the disadvantages, the report concluded:

"Public allocation or regulation is clearly necessary at some levels, particularly for intersectoral allocation. However, problems with this form of allocation are seen in poor performance of government-operated irrigation systems, leaking municipal water supply systems operated by public utilities, licensing irregularities and inadequate controls over industrial water use, and damage to fish and wildlife habitats."[9]

However, the report cites research, which argues "a major reason for such problems lies in the failure of the public allocation mechanism to create incentives for water users to conserve water and improve use efficiency."[10] To this one can add that, in Australia, structural adjustment payments from the federal government to successive state governments have been misappropriated into general state revenue coffers. A substantial part of these funds should have gone towards adjustment payments, or irrigation improvements, or water savings through state government infrastructure improvements.

On the other hand, arguably, Australian farmers are well down the road to implementing the necessary water efficiencies on their farms. They have been swift in taking up new irrigation methods and would have very likely invested even more on water savings infrastructure and technology over the past decade, but for the prolonged drought and declining farm incomes.

Also, some of the issues identified by the World Bank are less a problem in Australia than elsewhere in the world.

The World Bank report concluded:

"There is an essential role for the state (public allocation) in the development and management of water resources, particularly under circumstances involving large scale systems. The state's interest in many water resource investments relates to their strategic importance, e.g. because of its role in increasing food security or public health. In addition to such positive effects that may not fully be captured by the private users, negative externalities associated with much water use (e.g., downstream pollution) call for a strong regulatory role for the state. However, the resulting public allocation depends on the relative political influence of various stake holders."[11]

Because governments have focused the current water debate on open water trading to the highest bidder, governments have lost sight of the strategic role of low cost irrigation water in providing low cost, quality food to Australian consumers, and for export onto world markets corrupted by heavy subsidies. Costing irrigation water at the cost of delivery is vital for these social and economic objectives.

Further, governments seem to have lost sight of the strategic importance of low cost water to industries. Some heavy industry plants would be put out of business if they had to pay the same price as city users. "It has been calculated that aluminium smelters use 1.3 million litres of water to create a tonne of aluminium. If governments charged the smelter $2 a kilolitre, the cost would be more than $2,500 a tonne of aluminium and people would not be able to afford to buy a can of soft drink."[12]

Governments need to keep separate markets for water, just as they regulate separate electricity markets. Even though they draw on the same generators and use the same transmission lines, electricity markets are kept separate, with differential tariffs for households and businesses.

Misunderstanding the nature of water

The second pillar of the NWI is never stated. It is assumed that water is a private good. However, water is not a private good, but a "mixed good" with some important public sector aspects to its nature.

Economists speak of "private goods," like cars, houses, and white goods. These are produced and sold in open free markets.

At the other end of the spectrum there are "public

goods" like the police force, parks and gardens and environmental laws, which are provided by governments and cannot, or will not, be provided by the private sector.

In between are "mixed goods" that have some characteristics of public goods and some characteristics of private goods. Water is a mixed good. It can be treated as a public good provided by governments, for example in the building of reservoirs or the provision of environmental flows. Conversely, some aspects of water can be treated as a private good, for example, when a farmer trades some of his surplus water to another farmer, on a temporary basis, in the same catchment area. Clearly, water cannot be treated solely as a private good.

In the current debate over water, the NWI and others are treating water as a private good. This mistake arises, in part, because the NWI is under the Federal Government's National Competition Policy, which has been the driving force behind the deregulation of many sectors of the Australian economy, including the corporatisation and privatization of many former government instrumentalities. This has involved the reclassification of some former public goods, like railroads, as private goods, capable of now being provided by the private sector.

The fatal mistake of the NWI – and of the various bodies like the Productivity Commission and the Business Council of Australia – is that they assume that water can be treated primarily as a private good when it is a mixed good, with different characteristics in different markets and at different stages of collection, treatment and distribution. This is a fundamental misunderstanding of the nature of water.

Consider the following contradiction which arises from this misunderstanding. The Federal Government's NWI is attempting to treat water as a private good, when at the same time Canberra has allocated $500 million to buy 500 gigalitres of environmental water under The Living Murray proposal. Environmental water is always a public good! The government's NWI policy treats water as a private good, while in practice it is purchasing water as a public good. This illustrates a key flaw in NWI policy.

Australia: history of primary water allocation

Historically, Australian governments have separately allocated water for agriculture, urban, industry and environmental use. In one sense this happened informally, without governments ever needing to declare separate water markets in legislation. Still, it was well understood that there were separate markets. When Victorian Premier, Henry Bolte, made his famous statement that not one drop of water would come to Melbourne from north of the Great Dividing Range, he was stating that there were separate rural and urban water markets. It was a guarantee that water allocated from the Murray-Darling and Goulburn systems would be used for irrigation only and that it would not be tapped for metropolitan use. It gave farmers security of irrigation entitlements.

By and large, the water storages developed over the twentieth century for agriculture were separate from the urban storages, with limited overlap. For example, Snowy Mountain Authority water was intended primarily for irrigation, although a small amount was set aside for towns and industries and some went to supply Adelaide's water needs.

Again in the 1980s, separate water markets were retained when transferable water entitlements were introduced in NSW, South Australia, and then in Victoria and Queensland. Mostly, these involved only the limited transfer of sales water (temporary water) on a seasonal basis, although a small amount involved permanent water trade.

In the early 1990s, when COAG began working on the NWI, separate water markets were still assumed. Discussion on market-oriented water allocations saw water trade as being restricted to trading within the rural water market. Other efficiency gains were discussed within the context of separate water markets. For example, the 1993 COAG communique on water described the policy intentions on urban water use as reinforcing "the need for urban users to use water efficiently, for example by promoting water reuse and recycling, the adoption of more efficient technologies and by reviewing the effectiveness of pricing policies".[13] There was no discussion on buying rural water.

Even the 2004 COAG NWI agreement, the product of over a decade of work on water markets, made no mention of integrating rural and urban water markets.

The first mention of taking irrigation water for urban use was made in the Productivity Commission's *Review of National Competition Policy Reforms* in 2004.[14] It recommended to COAG that in future the NWI aim at

> "better integrating the rural and urban water reform agendas, including through facilitating water trading between rural and urban areas. [This was regarded as] probably the cheapest source of new water for several cities ..."

In one sense, from the time water entitlement was separated from property, making it tradable to anyone, then urban water authorities were in a position to purchase rural water entitlements. However, this "integration" of water markets was never the stated intention of COAG. Farmers were never told that water markets were going to be integrated. Rather, the push for integrated water markets came from the Productivity Commission. It stated:

> "While carrying forward the reform process begun under NCP [National Competition Policy], the NWI and Murray-Darling water agreement do not exhaust all of the potential reform opportunities in the water sector.

> "The Commission observes, for example, that little attention appears to have been given to the scope to better integrate the rural and urban water reform agendas ... In the Commission's view, if water is to be allocated to its highest value use in the future, the urban and rural water markets will need to become increasingly integrated ...

Wider application of this approach could potentially reduce the need for the sort of prescriptive demand management approaches that have become increasingly common in urban areas in recent years."[15]

What the Productivity Commission initiated in its review of National Competition Policy is now a major focus of the Commission's policy recommendations, of the Business Council of Australia's water agenda and of the NWI.

This goes far beyond any of the earlier COAG or NWI communiques and water policies.

Why is there now a proposal to integrate rural and urban water markets when this was not stated in COAG's NWI? Why is this being proposed when both the FAO and the World Bank research shows clearly that primary water allocations among sectors cannot be left to market mechanisms? They both regard that, because water is a mixed good with many "public good" characteristics, the primary allocation of water among various markets has to be a decision made by the government. Why is there no discussion on creating new reservoirs, given the abundant availability of untapped usable fresh surface water, particularly in Queensland, NSW and Victoria?

Conclusion

There are long discussions, over many pages, in several Productivity Commission reports attempting to grapple with what it recognises are numerous problems of attempting to create a single water market. Unfortunately, while it recognises that there are many problems, the Commission fails to provide practical and convincing solutions. It is attempting to treat water as a "private good" when water is a "mixed good".

The first object of water trading – to allow open markets to allocate water among agricultural, urban, industrial and environmental uses – is a fatally flawed concept.

It cannot be done, and attempts to allocate water using open market trading will cause major economic and social disruption.

> **Recommendation:** Governments must ensure that the water markets for agriculture, urban, industrial and environmental uses remain separate markets, with separate prices based on delivery costs. Governments must recognise that water is a mixed good, not a private good, with many public good characteristics. Because water is a mixed good, with different characteristics in different markets, only governments can decide *primary water allocations* among these markets. It cannot be allocated through a single, open trading market, selling water to the highest bidder. Particularly given the current chronic drought, governments must immediately quarantine irrigation water so that it cannot be permanently traded for urban, environmental or other uses, so as to ensure security of supply to the agricultural sector.

Endnotes

1 *Rural Water Use and the Environment: The Role of Market Mechanisms: Discussion Draft*, Productivity Commission. 2006, p. 212.

2 Ibid., p 248.

3 Turner, Kerry; Georgiou, Stavros; Clark, Rebecca; Brouwer, Roy; *Economic valuation of water resources in agriculture* (Centre for Social and Economic Research on the Global Environment; Zuckerman Institute for Connective Environmental Research; University of East Anglia, Norwich; United Kingdom of Great Britain and Northern Ireland); for FAO, Rome 2004. Chapter 3. "Economics of Water Allocation". http://www.fao.org/docrep/007/y5582e/y5582e00.htm#Contents

4 Dinar, Ariel and Rosegrant, Mark W., *Water Allocation Mechanisms: Principles and Examples* (World Bank, Agriculture and Natural Resources Department; and International Food Policy Research Institute), World Bank 1997, p. 8-9. http://ideas.repec.org/p/wbk/wbrwps/1779.html

5 Koehler, C. L. 1995. Water rights and the Public Trust Doctrine: Resolution of the Mono Lake controversy. *Ecological Law Quarterly* 22(3): p. 541-590.

6 Moench, M. 1995. Allocating the common heritage: Debates over water rights and governance structures in India. National Heritage Institute, San Francisco (mimeo).

7 Turner, Kerry; FAO report, Op.cit., Chapter 3.

8 Dinar, Ariel; et. al., World Bank report, Op.cit., p. 3

9 Ibid., p. 11.

10 Ibid., p. 11.

11 Ibid., p. 32.

12 McManus, Gerard, "But never a drop to drink", *Herald Sun,* September 28, 2006.

13 Council of Australian Governments, Communique, 29 August 2003.

14 *Review of National Competition Policy Reforms* (No 33, 2005), Productivity Commission, p. 209 & 206.

15 Ibid., p. 203-204.

4

Trading water from "low value" to "high value" agriculture

The third pillar of the National Water Initiative (NWI) water trading policy was to allow the trading of water from low value to high value agriculture. This concept is fundamentally flawed. Economic modelling claims to show substantial benefits of allowing water trade for this purpose. These modelling outcomes are dubious, for the following reasons.

1. Defining low and high value agriculture

What is classed as low value as opposed to high value agriculture? As the Australian Farm Institute points out, agriculture is worth 3.2 per cent of the economy, but the farm dependent economy – the input sector, agriculture, and the output sector – make up 12.2 per cent of the economy, valued at about $72 billion.[1] In other words, the value of a particular farm product is not measured just by the farm gate price, but by the added value from farm inputs and outputs.

Hence, a high farm gate price product may be worth

less to the economy than a low farm gate price product. If there was water trade to the high farm gate priced product, it would incur a net cost to the economy. In this case, water trade involves a market failure.

2. High value today, low value tomorrow!

Only about 12 per cent of Australia's agriculture is considered "high value". Hence there is only a limited amount of water that can be taken from "low value" for "high value" agriculture, without leading to overproduction. Take the wine grape industry. Substantial amounts of water have been traded to corporate wine grape farms that are financed by large managed investment schemes, yielding substantial tax breaks to investors. This has led to a significant oversupply of a number of wine grape varieties, with prices falling from $600-800/tonne to $150-200/tonne in 2006.

Again this raises a problem of defining what is a "high value" farm industry. What is today's "high value" industry can easily be tomorrow's "low value" industry. What validity is to be placed on economic modelling showing big increases in agricultural production from water trade, if the industries concerned can rapidly become low value industries because of over investment? Agricultural history is full of boom and bust stories, reducing highly optimistic modelling figures to mere guesstimates.

3. The importance of low value farming

Conversely, 88 per cent of Australian agriculture is "low value", as measured at the farm gate. This can be taken two ways.

Some of this becomes "high value" in output industries, adding substantially to the economy, and to exports.

Some of this goes on to provide low cost food and fibre products, with two desirable objectives:
• providing low cost, quality food products to Australian consumers; and

- giving Australian farmers a competitive edge on world markets for bulk commodities like grains.

    Hence, it is a mistake to treat what is classed as "low value" agriculture as second class to so called "high value" agriculture. Each has legitimate but different economic and social objectives.

4. High value can also be high cost

From the viewpoint of farmers, they want to make a profit. High value agriculture may or may not provide high profits, because high value often means incurring high input costs. Hence, it does not automatically follow that high value means high profits and a reasonable return on investment for farmers.

5. Factoring in transmission losses

The further water is traded out of a catchment, the more are the transmission losses, and the greater the costs from the losses.

    The corollary is that if more water is used in the catchment area, then less is lost in transmission and more water is available for use and for adding value to the agricultural economy.

6. Shift agriculture not water

Seldom understood is the fact that, not only is there a limited amount of high value agriculture, but there is no need to shift water out of catchment areas from low to high value agriculture. The nature of the Murray-Darling and Goulburn systems is that so called "high value" products can be grown almost anywhere in this large region. The climate, soil types, and other conditions are common right across the basin. Therefore, it is better to shift agriculture to where the water is available, rather than shift the water to where a particular farm is situated.

    Furthermore, this is what farmers are already doing.

The Productivity Commission says that farmers are well informed about production alternatives and irrigation technology. In part, this is why, as its own research shows, agriculture is Australia's third highest productivity industry, after the communications and wholesale sectors.[2] Farmers change their production to maximise the value of their output.

In short, agriculture already shifts to where water is available, negating the need for water to be traded out of a catchment area. There is no need to have water trade to boost agricultural output.

7. Physical restraints on water trade

There are other practical impediments to trading and transferring water long distances:

- only a limited amount of water can physically be traded between regions because of local needs and channel flow restraints; in many regions, not much more can be traded than what is being now traded; and

- there are geographical constraints on water trading – Darling River water that enters the Murray at Wentworth will not flow back up the Murray to Cobram and down to Shepparton.

8. Environmental restraints

There are good environmental reasons to restrict water trading between catchments. Over the past 80 years of irrigation in the Murray Darling Basin, farmers, scientists and catchment management authorities have developed strict rules on how much water can be used on land in the various regions without creating problems like rising water tables and salinity.

Open water trading would put at risk the environmentally complex array of irrigation rules across many catchments and risk environmental degradation of farmlands.

9. The full cost of water trade

How does one conduct a true cost-benefit analysis of a water trade from "low value" to "high value" agriculture? The following seem to be important factors and issues somewhat neglected by policy makers.

- How much water is lost in transmission when water is traded over long distances between catchments?
- What are the benefits of water trade to input, agriculture and output industries in farm areas gaining water? What are the costs to input, agriculture and output industries in farm areas losing water?
- Given that irrigation is geared not just to a particular farm, but to all farmers along an irrigation channel, then what is the full cost to farmers and water supply authorities resulting from the stranding of assets?
- What are the economic losses and gains in local communities when water is traded from one region to another?

10. Water and financial security

Banks lend for 10-year farm investments, with the farmer's land title and water right as security. Should water rights be detached from property titles, allowing farmers to sell off their water, then banks are likely to curb lending for farm investment; or they will insist on holding the water title as security. As the risk of stranded assets increases, this will make it more difficult for banks to decide on lending to those who can easily become third party victims of water trade.

Conclusion

In stark contrast to the belief that water trade can boost agricultural production, this chapter shows that it is entirely unnecessary to unbundle water entitlements from land titles, in order to create water trading markets, so as to shift water from low value to high value crops. The concept may appeal to theoretical economists, but it is fatally flawed as a means of boosting agricultural output. As is so common in agricultural industries, today's high value product becomes subject to over investment, made worse by managed investment schemes, resulting in tomorrow's low value product.

Those who predominate in this debate fail to understand the economic and social objectives of low value products, both to Australian consumers and Australia's trading position in the world.

Further, the water being traded is more likely to go to managed investment schemes or to environmental flows, such as for the Living Murray. In which case it is highly unlikely that this water will return to collapsing irrigation areas.

Finally, to the extent that irrigation water will be lost to other uses, the shortage of water in a tight market will substantially force up the price of irrigation water.

Recommendation: Governments should recognise that the idea of trading water from low value to high value agriculture is a fundamentally flawed concept. The final value of product to the economy depends on down stream processing, not merely farm gate price. What is a high value product today, can quickly become tomorrow's low value product, as is currently happening in the wine grape industry. Further, excessive emphasis on "high value" farm gate price products conflicts with other economic and social objectives, like providing the Australian people with low cost food, and helping Australian farmers achieve a competitive advantage over their heavily subsidised foreign competitors. Governments must recognise that it is better to shift agriculture to available water than to shift water to agriculture; that this is the common practice of farmers; and that farmers are in the best position to judge water use for "low-value" and "high-value" crops.

Recommendation: Governments must ban permanent water trade out of catchment areas. Trade of
secure irrigation water should be restricted to trade
among farmers within a catchment area.

Endnotes

1 *Australia's Farm Dependent Economy: Analysis of the
Role of Agriculture in the Australian Economy,* research paper by the Australian Farm Institute, 2005.

2 Banks, Gary, Productivity Commission Chairman, *The
drivers of Australia's productivity surge.* Paper Presented at Outlook 2002, hosted by the Department
of Industry, Tourism and Resources and the Australian Bureau of Agriculture and Resource Economics,
National Convention Centre, Canberra, 7 March.
Figure 4, p. 6.

5

Effects of Water Trading

Water in farmers' cost structure

Open water markets will push up the price of irrigation water.

The Productivity Commission (PC) argues that any price rises from anticipated rural to urban water trade, and from other price pressures associated with deregulation of the water markets, can be absorbed by farmers.

> "Water is not a large component of most agricultural input costs. In the most water-intensive irrigated industries, such as rice growing, the cost share is 10–20 per cent. In irrigated industries where capital and labour intensity is higher, such as horticulture, water's share of input costs may be in the range of 1–2 per cent (Appels, Douglas and Dwyer 2004[1])."[2]

Logically, if this statement is true, then raising the price of water will not be much of a punitive incentive to ensure water savings measures. In reality, the claim is simplistic and indicates the need for more detailed knowledge of farm industries and variability of circumstances.

The claim that water costs make up only 1-2 percent

of costs in horticulture depends on whether this includes the cost of on-farm enhancements, such as a drip irrigation system, small sprinklers, overhead sprinklers, mobile sprinklers, filtering systems and power costs for pumping. Factored in, horticulture water costs are more likely to be 3-5 per cent of costs.

Consider other forms of farming, such as a Victorian dairy farm with, say, a 300 megalitre water title. It was not long ago that the cost of delivery was $30 per megalitre. It then rose to about $60 per megalitre. It is anticipated to rise to $100 per megalitre in the near future. This represents cost rises respectively from $9,000 to $18,000 to $30,000. A rise from $30 to $100 would wipe out the profit margin of many dairy farmers.

Further, in drought years the price of water can increase dramatically. During recent drought years, farmers in different industries were reporting water costs of 30 per cent to 50 per cent of their farm costs.

Some forms of farming are better able to absorb water cost increases than others. Horticulture in central Victoria can absorb higher cost increases than dairy farmers. On the other hand, horticulturists are forced to absorb very high water costs in drought years. If an apple tree is starved of water for one year, it will produce no fruit in the next, and the farmer will lose two years of production, even if there is plenty of water available in the second year. Such a farmer is forced to pay a high price for water in order to keep his farm going. Should the drought extend for two years, it will be all the more costly.

Dairy farmers cannot cover costs if they have to pay the high price that horticulture can pay in drought times. Alternatively, if water is cut from a dairy farm, then farm production drops proportionally. Dairy farmers cull their herds, which take several years to rebuild. It also takes some years to rebuild good pastures.

Other issues also come into play. Horticulture largely goes to the domestic market. Victorian dairy produce is largely for export. The exported product is significantly valued added, positively assisting Australia's large trade imbalance and current account deficit.

The PC, NWI and Business Council of Australia water policies overemphasise the savings from drip and spray irrigation over flood irrigation. Which of these are the more efficient delivery systems depends on a number of factors, such as soil types and absorption characteristics, crop type and land gradients. Then there are the added costs of various delivery systems, in terms of infrastructure, maintenance and pumping costs.

In many cases, flood irrigation is still the most efficient, and the lowest cost, delivery system. There is good evidence to show that, in the right circumstances, well managed flood irrigation is more efficient than poorly managed spray irrigation systems. In other cases, the highest output yield comes from a combination of drip, spray and flood irrigation at different stages in the growing season. The costs are high, but these multiple delivery systems are used to achieve high value crops, largely for export.

The economics of delivery costs depend also on the anticipated farm gate return, which depends on the type of crop and quality grade target. Consequently, there is no single, efficient water delivery system. It depends on local circumstances.

Recommendation: Governments must recognise that there is no single ideal form of on-farm irrigation delivery system and that farmers should be provided financial incentives to invest in the most efficient forms of irrigation for their particular farm products.

Recommendation: Rather than using price penalties to encourage water savings on farms, governments should offer positive financial incentives to farmers for on farm enhanced water savings and intensification of agricultural land use, including where this results in parts of farmland being transferred from marginally profitable farming to being used for environmental purposes. Positive incentives should be offered, just as the Federal Government provides financial incentives to the states to implement competition policies under National Competition Policy.

What are water rights?

"Water entitlement" (or "water allocation", or "permanent water", or "secure water") refers to a farmers' annual entitlement to a set amount of water with a particular level of security and a particular reliability of delivery over 100 years. This can vary between districts or irrigation systems depending on the local characteristics of water resources and local system infrastructure. Water entitlement was traditionally attached to a farmer's property title, and while the water could be traded on a seasonal basis, the right could not be detached from the land and sold permanently.

However, over the past 20 years, governments have moved to detach farmers' water rights from their property title and to make secure water tradable on a limited basis. The NWI now proposes unlimited water trade.

The price of water right, or high security water, was set at the cost of delivery, plus a service fee in the district of origin. The latter covered the management and maintenance of infrastructure. Now, irrigation districts are imposing substantial exit fees on water permanently traded out of their system, in order to compensate for loss of water and its use to the irrigation district.

"Sales water" refers to water that Victorian farmers can purchase on a seasonal basis over and above their permanent water entitlement, depending on annual availability. Sales water can only be made available after permanent water has been fully allocated. Sales water has always been tradeable within and between irrigation regions. The price is set by a complex formula.

Under the NWI, water trading will allow farmers to sell off their permanent or secure water to any buyer. In Victoria, there is one restriction. No one can hold more than

10 per cent of an irrigation region's water right without being attached to land.

Water trade: conversion issues

"Oils ain't oils", and "water ain't water". Just as different oils have different physical characteristics, water has different entitlement, reliability and delivery characteristics in different markets. In fact, there are 24 different types of water entitlements in the Murray Darling Basin with different characteristics, security levels and resources to back up appropriate delivery entitlements. Most policy makers do not understand the complexity of water titles, and all state governments are yet to undertake audits of these entitlements. Without an understating of these different entitlements, how can water trading take place? How can farmers in different irrigation regions know what they are buying?

The differing water entitlements are not the result of the states and water authorities opportunistically restricting water markets to local farmers at the expense of all other considerations. Rather, the differing entitlement systems were planned carefully, based on regularity of supply, size of water reserves to back supplies, nature of delivery systems and types of soil. In turn, these factors helped determine the types of agriculture predominating in different regions. The differences in water entitlements reflect the differing natural features of regions. There was a great deal of long-term foresight in the planning and creation of different water entitlements systems.

For example, one megalitre of general security water in southern NSW has different levels of security and resources to back up delivery entitlements when compared to one megalitre of secure water entitlement in the Goulburn Valley of Victoria. The availability of high security water with a high reliability of supply is necessary for fruit trees in the Goulburn Valley, because the loss of trees due to drought means that replantings will take seven years to mature and bear fruit. In contrast, in another irrigation area producing various vegetables and grains, they can farm on lower security water. After a drought, they can be replanted and produce a crop as soon as water is available again.

In order to try and make a water trading system work, attempts have been made to create water exchange rates between different irrigation regions. One attempt at an exchange rate is water "tagging", where water retains its original source reliability characteristics from it source. This sound fine in theory, but there are problems. For example, if water is traded from NSW to Victoria, and NSW only has a 30 per cent allocation this year, will the Victorian farmer receive only 30 per cent of the water for which he has paid, or will the farmer receive his full entitlement? If he gets his full entitlement, where will the extra water come from?

Another problem is that water is lost in transmission. If a Victorian farmer buys 100 megalitres of water from NSW and 25 megalitres is lost in transmission, will the buyer receive 75 megalitres or 100 megalitres? If he receives his full allocation, where will the extra 25 megalitres that was lost in transmission come from?

These are but a sample of the numerous problems in trying to convert water from being a mixed good to a tradable private good. It will most likely prove impossible. It is a fundamentally flawed concept.

Water trade requires the unbundling of a complex system of water entitlements that vary widely among regions and states. To take a legal analogy, the deregulation of water trade is comparable to the abandoning of the whole system of common law and replacing it with a bill of rights. Such a momentous change to the legal system would involve major public debate and a long period of deliberation about which was the best system. What is extraordinary in the water debate is the lack of farmer representation and the ignorance of so many professionals on water issues. Many so-called "experts" come from theoretical economic backgrounds and seem bent on trying to make water fit a free market model, rather than to understand the nature of water and the practical issues involved in irrigation farming.

Water's changing supply and demand

It is likely that for some years into the future, the supply of water for human use will be relatively short of demand, for various reasons:

- Australia appears to be going through one of its periodic, cyclical dry spells.
- agricultural managed investment schemes, with big tax write-offs, have invested heavily in tree plantations. The latter are large in size, and many are in the catchment areas of major reservoirs, or rivers and streams used for irrigation. These plantations are absorbing large amounts of naturally falling rainwater, substantially reducing runoffs and therefore, the availability of water for urban and other agricultural use.
- the destructive Alpine fires of recent years, resulting from bad national park management, have permanently altered the region's hydrology, increasing stream flows in the short term, but probably reducing flows in the medium-to-long term.
- poor land use management in some catchments, where hobby farms with substantial onsite dam capacity have reduced flows into major water storages.
- the "no new reservoirs" policy adopted by several state governments.

On the other hand, demand for water is likely to rise, because:

- in dry periods there is a greater demand for water.
- increasing demand for environmental flows, like the 500 gigalitre flow for the Murray River, will reduce water availability for other uses.
- over the next 50 years, Australia's population is expected to expand by several million, bringing urban demands into increased conflict with irrigators.

- state governments are hooked on the revenues that come from the rapid expansion of metropolitan areas, where land values are high and where urban water supplies are tight. Instead, governments should be encouraging population expansion in areas where there is plenty of available water, like the north coast of NSW and central and northern Queensland.

- managed investment schemes involving large scale farms, particularly in forestry and wine grapes, are distorting the water market. (See later).

There seems to be an attitude amongst policy makers that, as farmers use 66.9 per cent of all water consumed in Australia,[3] taking a small amount for urban and other uses will not seriously hurt irrigation agriculture. This is false.

As the price curve for water is inelastic, in a tight water market small increases in the demand for, or a decline in the supply of, water causes a sharp increase in the price of water. This is what happened to irrigation water prices during the recent drought.

The consequences of water trade

Open water trade will invariably force up the price of irrigation water. There are a number of ways in which this can occur and with predictable consequences.

1. Water barons

The problem of price spikes is likely to increase if non-users buy up substantial amounts of water rights. Open water markets allow water barons to buy up water. This gives them the ability to manipulate the water supply in a tight market, pushing prices to very high levels, particularly in times of drought. This can also happen in some electricity markets. Witness the manipulation of electricity prices in California to extraordinary high levels by companies controlling only 6-8 per cent of the market during the 2000-2001 electricity crisis.[4]

2. Cities will always outbid farmers

In open water markets, cities will always be able to pay more for water than farmers. Urban authorities can spread the cost over businesses and households, where marginal price rises are more easily absorbed. On the other hand, farmers are price takers, not price makers. They cannot easily pass on cost increases.

Urban areas will invariably win out over irrigation users. Recently, the Victorian government has proposed a pipeline to take 38,000 megalitres of water from the Goulburn Valley irrigation system for Bendigo and Ballarat. During the recent debate over the failed Snowy Hydro public float, proposals were aired to pipe water, set aside for irrigation farming, from the Snowy to Sydney. In 1967, Henry Bolte's government promised that the new Upper Thomson Dam would provide 144,000 megalitres annually to Gippsland farmers. A decade later, a different government reduced this pledge to 12,000-14,000 megalitres annually to irrigation farmers. The most ever delivered was 10,000 megalitres, and the rest has been piped to Melbourne.

The Business Council of Australia has endorsed similar proposals, saying:

> "As Ross Young, the Executive Director of the Water Services Association of Australia (WSAA) has said: 'What is often forgotten is that with the exception of Sydney, all the other capital cities share water supplies with agriculture users which enable water to be transferred without the need for any new infrastructure. In the long term, the prospect of Sydney being connected to the Snowy Mountains scheme should not be completely ruled out. Melbourne already shares water with irrigation from the Thomson dam and only a short pipeline is required to connect Melbourne to the Goulburn system to open many trading opportunities across Victoria.'

> "It is clear that Adelaide, Melbourne, Perth and Hobart already have ready access to rural water.

Brisbane can also gain easy access, although expensive pumping would be required rather than relying on gravity."[5]

3. Stranded assets

As the price of water increases, some farmers will go out of business and sell off their water rights.

As some leave, this will have the effect of stranding assets. Along any irrigation spur channel there is a percentage loss of water. A typical, earthen, unlined channel will operate on a 75 per cent efficiency rate. This means it will lose 25 per cent of the water through seepage and evaporation in the delivery process. If only a small number of farmers sell their water out of the channel, the efficiency rate can drop to 50 per cent. With half the water lost in delivery, the cost of delivery makes the channel uneconomic.

At that point, all the farmers go out of business, whether they are efficient or not. They may be highly efficient producers and profitable farmers, but with the loss of their water delivery, their land and farm assets fall substantially in value.

This produces stranded assets, which can include dams and diversion works; major channels and diversion infrastructure; local channel and delivery works; on-farm irrigation delivery systems; and other on-farm infrastructure assets associated with irrigation activity.[6]

The Productivity Commission (PC) says that simulated outcomes from the loss of 10 per cent in irrigation water from a region showed that the losses were "unlikely to have significant consequences for regional economies."[7] The Commission also cites research by the CSIRO which argued that there was no evidence that water trading had caused "land values to plummet", or left supply authorities with "unserviceable infrastructure debts", or led to a "decline in the region's economic activity."[8]

It appears that thorough research measuring the actual effects of irrigation losses on regional economies is yet to be done. However, Mildura region farmer Danny Lee has examined the broad effect of water trade out of irrigation areas. Unpublished figures available at Goulburn Murray Water show that from 1994 to June 30, 2006, net 150,000 megalitres of water right was traded out of northern Victorian irrigation regions,[9] mostly to managed investment schemes between Nyah and Mildura.

In 2001, the Victorian Department of Natural Resources and Environment estimated the value of irrigation water to regional communities.[10] From this report, it can be calculated that:

- the farm gate product value from one megalitre of water was about $1,000 (an average figure across various industries);
- the flow on effect in the regional community from each farm gate dollar was 4-to-1 (i.e. a multiplier of 4); and
- the farm capital write down for each megalitre of water lost from a farm was about $1,400, on average.

Consequently, the cost of water trade can be estimated. The loss of 150,000 megalitres of irrigation water from northern Victoria has resulted in:

- farm gate losses of $150 million annually;
- the loss of $600 million from the regional communities annually; and
- a total write down of $210 million in farm assets.

It could be argued that some areas suffering losses are losing only relatively small amounts of water. The problem is that, as the demand for water is set to grow while the supply remains below average, more water is set to be lost from irrigation areas, having a cumulative effect on irrigation regions. The current severe drought has seen some farmers selling their permanent water entitlement, in the hope of obtaining enough cash to survive the season, and then buy their water back in a year or so. In August 2006, there was a ten-fold increase in the amount of water traded in parts of the lower NSW irrigation areas.[11] If secure water is sold out of these areas to managed investment schemes or for Living Murray environmental flows, this water will not come back into these irrigation regions.

This indicates that the full costs of water trade out of a channel and a district will be significantly higher than suggested by the PC in its various reports, including its *Stranded Irrigation Assets*[12] working paper. Seeing Federal structural adjustment payments – that should have been used for regional readjustment – misappropriated into the states' general revenue coffers, only rubs salt into the wounds of people in these regions.

Indeed the NWI appears to have partly recognised the damaging effect of water trade. It limited annual trade out of any irrigation region to 4 per cent of the total water entitlements of that area, subject to review in 2009 with a move to full and open water trade in 2014. Further to this, the NWI agreed that irrigation authorities could impose cost restrictions on the outward trade of water from these regions. The restrictions on outward trade include:

- ongoing payment of infrastructure access fees by landowners after the sale of an entitlement;
- "tagging" – access fees set in the source area are paid by the new owner of an entitlement (along with the access charge in the destination area); and
- "exit" fees paid to the infrastructure operator by the seller or purchaser on the sale of an entitlement.[13]

The Business Council of Australia and the PC have criticised the NWI for permitting these restrictions on water trade among irrigation regions. They argue that the restrictions "will reduce the economic gains potentially available from entitlement trading".[14] The economic gains are supposed to come from trading irrigation water from "low value" to "high value" agriculture. The PC extends their argument to the point of saying, "Indeed, the existence of stranded assets can indicate that the market is working to redistribute entitlements efficiently."[15]

What the PC, Business Council and NWI have all failed to acknowledge is that the bulk of water is being traded to managed investment schemes (MIS).

4. Managed investment schemes (MIS)

MIS behaviour in the water markets is critical to understanding what is really happening in water trade among irrigation regions.

Investment in MIS is not based on economic forces that lead to the most efficient allocation of resources, but on large up-front tax concessions to Collins Street investors. This causes over investment in various areas of agriculture, turning today's high value farm products into tomorrow's low value products, and putting many genuine farmers out of business. This is happening in wine grapes, olives, mangoes and a number of other rural industries.

Consequently, economic modelling, which claims to show that water trade to high value products will boost the total value of agricultural output, appears to be based not on market forces generating water transfers between irrigation regions, but on huge investor tax breaks for MIS. (See Chapter 6 for a more detailed discussion of MIS).

In turn, the problem of stranded assets is not the result of water trade to high value products and "a sign that the market is working to redistribute [water] entitlements efficiently", as the PC claims; rather, stranded assets are due to over investment in MIS. These might produce high value products today, but they are set to be low value in the near future.

Interestingly, the PC has side stepped the critical issue of MIS. In the Commission's final report on water trading and the environment last August, it discussed MIS. However, the report claimed that it was "beyond the scope of the present study to determine whether MIS and related tax arrangements have a net positive or negative impact on the community as a whole (or on rural water use specifically), or whether they constitute a 'concession' (or subsidy) to particular industries, businesses or individuals."[16]

One can only speculate as to the reason for reaching this conclusion in relation to water, especially as one submission to the Commission pointed out that most water trade in agriculture was to MIS.[17]

Perhaps it could be argued that to accept that the bulk of water trade to MIS is for tax reasons, and not to high valued agriculture for economic efficiency reasons, destroys the whole rationale for water trade to high valued agriculture; it destroys the second pillar of the NWI water trading policy!

Perhaps to advise the Federal Government of such a conclusion would have been, to borrow from Sir Humphrey Appleby, "a courageous decision".

5. Environmental flows

In an open water market, governments purchasing environmental flows are likely to win out over irrigation users, if governments continue to listen to environmental advocates in place of sound scientific evidence. Currently the Federal government is proceeding with an allocation of $500 million to buy 500,000 megalitres of water for environmental flows down the Murray River.

If the government were to buy this water directly from farmers, then the price of irrigation water would be forced up, especially in this relatively dry period. Partially aware of this problem, in April 2006 the Federal parliamentary secretary for water, Malcolm Turnbull, unveiled plans to buy irrigators' water for the environment, but only if it comes from water savings from farmers achieving greater water efficiencies.

This proposal is unrealistic.

First, because Australia is in the grip of a long dry period, farmers are facing a decline in their water resources. Any savings from more efficient water technologies is likely to be kept on the farm, to compensate for the reduction in irrigation water supplies.

Second, installing new on farm water savings technologies costs from $4,000-5,000 per megalitre. Given the Federal Government has only allocated $500 million to buy 500,000 megalitres, this works out to be an offer of only $1,000 per megalitre for the water. Farmers are not going to sell water for $1,000 when it is costing them 4-5 times as much from on farm investments.

Again the assumption seems to be that, because farmers use two-thirds of Australia's annual water consumption, taking some of this for environmental flows won't damage agriculture. This is wrong. In a tight water market, taking water for environmental flows will force up the price of irrigation water.

An even more fundamental question about the Murray River environmental flows has to be asked. Are the flows even needed?

The House of Representatives Standing Committee on Agriculture, Fisheries and Forestry undertook the first independent evaluation of the science behind the Living Murray proposal for environmental flows in 2004. Its members were aghast at the poor quality of scientific evidence presented.

The committee issued an interim report in which it urged that no environmental flows be committed to the Murray River until a full scientific study was done scoping all the issues affecting river health. It recommended that a slice of the $500 million allocated for environmental flows should be reallocated to ensure the scientific work is undertaken thoroughly. Governments have ignored these recommendations. (See Chapter 7 for a longer discussion of environmental flows).

Conclusion

The price curve for water is inelastic. If Australia is suffering from below average precipitation and rising demand for water, then in an open market for water the price of water will substantially.

The NWI is supposed to address the water shortage issue. Instead, its water trade policy is set to price many farmers out of business, causing large write downs of farm assets. The process won't happen overnight, but the attrition of farmers from higher value, intensive irrigation farming will progress at a steady pace.

Should this scenario continue, it must bring into question the long-term economic sustainability of substantial areas of irrigation agriculture. This could see Australia being increasingly left reliant on dry land farming.

Recommendation: The Federal government should provide financial incentives for the states to undertake a full audit of irrigation entitlements, particularly ground water entitlements.

Recommendation: The Federal Government must hold the states accountable for National Water Initiative payments, ensuring that these payments are used in water savings infrastructure and/or for the construction of new water storages, and not diverted into general revenue.

Recommendation: Governments should develop new water infrastructures in regions where there is ample water supply. Population migration should be promoted by grants and assistance for business transition arrangements so that skills in irrigated farming and irrigation support services are moved to areas where expansion of irrigation is possible. This would reduce pressures on regions with insufficient water supply.

Endnotes

1 Appels, D, Douglas R and Dwyer, G 2004, *Responsiveness of demand for irrigation water: a focus on the southern Murray-Darling Basin*, Productivity Commission Staff Working Paper, Canberra. Cited in Productivity Commission *Discussion Draft,* p. 210.

2 Op.cit., Productivity Commission *Discussion Draft,* p. 210.

3 *Rural Water Use and the Environment: the Role of Market Mechanisms,* Productivity Commission's Final Report, August 11, 2006, p. 264.

4 Borenstein, Severin, "The Trouble With Electricity Markets: Understanding California's Restructuring Disaster", *Journal of Economic Perspectives*, Volume 16, Number 1, Winter 2002.

5 *Water Under Pressure: Australia's man made water scarcity and how to fix it,* Business Council of Australia, September 2006, p. 27.

6 *Rural Water Use and the Environment: The Role of Market Mechanisms,* Productivity Commission's final report, August 11, 2006, p. 92.

7 Peterson, D., Dwyer, G., Appels, D. and Fry, J., *Modelling Water Trade in the Southern Murray–Darling Basin*, Productivity Commission Staff Working Paper, Melbourne, November 2004. Cited in *Stranded Irrigation Assets: Staff Working Paper,* Productivity Commission, June 2006, p. XII

8 CSIRO Land and Water, *Inter-State Water Trading: A Two Year Review*, Draft Final Report prepared by Mike Young, Darla Hatton MacDonald, Randy Stringer and Henning Bjornlund, December 2000. Cited in *Stranded Irrigation Assets: Staff Working Paper,* Productivity Commission, June 2006, p. 37.

9 Goulburn-Murray Water, unpublished figures on the annual trade of secure water in and out of northern Victoria's irrigation regions from 1994 to June 30, 2006. Data supplied by Danny Lee.

10 *The Value of Water: a guide to water trading in Victoria,* Department of Natural Resources and Environment, in conjunction with rural water authorities, December 2001.

11 ABC *AM Program,* September 22, 2006.

12 *Stranded Irrigation Assets: Staff Working Paper,* Productivity Commission, June 2006.

13 Ibid., p. XII.

14 Ibid., p. X. Also see, *Water Under Pressure: Australia's Man-Made Water Crisis and How to Fix it,* Business Council of Australia, 2006, p. 47.

15 *Rural Water Use and the Environment: The Role of Market Mechanisms,* Productivity Commission's final report, August 11, 2006, p. X.

16 Ibid., p. 133.

17 Byrne, Patrick; O'Brien, John; Eagle, Neil; McDonald, Neil; Submission to the Productivity Commission in response to the Rural Water Use and the Environment: The Role of Market Mechanisms Discussion Draft, July 13, 2006.

6

Managed investment schemes: distorting water allocations

Managed investment schemes (MIS) are distorting rural investment, agricultural markets and water allocations. MIS involve wealthy investors receiving large up front tax breaks for investments in large corporate farms. They are distorting water markets, forcing up the price of irrigation water in several ways:

- By charging investors several times the cost of establishing the project, MIS are financially able to buy up large amounts of water entitlements above normal market prices, making them responsible for the bulk of water traded between catchments.

- MIS tree plantations in catchment areas are absorbing water at zero cost, reducing surface and ground water flows into streams, rivers and water storages at considerable cost to other irrigation water users.

- At the end of the MIS, after say fifteen years in the case of blue gum plantations, the manager operators are left with huge water banks, making them water barons and giving them the ability to manipulate water market prices.

About $3.6 billion has been invested in MIS since the 2001-02 financial year, with almost $1.2 billion spent in 2006. They have an annual compound growth of around 3 per cent,[1] involving the purchase of about 105,000 hectares of land annually.

Of the 59 new projects in 2006, investments break down as follows: 44 per cent into tree plantations, 42 per cent into horticulture, 5 per cent into various agricultural investments and 9 per cent to other investments. The type of projects of recent years include: forest planting, "tax-effective tomatoes ripening in giant glasshouses in northern NSW, pearls in the Northern Territory, sandalwood, vineyards, olives, mangoes, almonds, truffles, walnuts and cherries."[2]

In part, these schemes operate by charging a high up front fee, several times the cost of establishing the project. For example, a tree-planting project can see investors charged as much as $9,000/ha, while the true establishment cost may be only $1,500/ha. Without going into the details of the tax minimising aspects of these MIS, the point is that with tax minimising schemes,

> "most investors don't worry about a return at the end of 15 years. Their main concern is a tax deduction now. Given it is a 100 per cent deduction, the more they spend the better it is. And with any profit years away, it is unlikely the promoter will be held responsible for that performance until too late."[3]

Hence, MIS are not based on market signals, on laws of supply and demand, or on issues of efficient allocation of land and water resources. They ignore the market signals that are supposed to ensure resources are allocated "to the highest use ... [so] that the community gets the greatest return (broadly defined) from its scarce resources", which as the Productivity Commission (PC) says is the efficient way to allocate resources like water. They are solely driven by wealthy investors – many having made a lot of money out of the recent bull-run on the Australian stock market – wanting to minimise their tax. The schemes are tax driven.

MIS frequently lead to overproduction and the collapse of rural commodity prices. The price collapse puts other farmers out of business, but it does not affect the operation of the MIS, as the primary purpose is to minimise investors' tax liabilities, not to actually make a profit. This overproduction is rapidly turning some high value farm products into low value products. Responsible for 15 per cent of the wine grape industry, MIS have contributed to the collapse in prices for wine grapes, from around $600-$800/tonne a few years ago, to $150-$200/tonne in 2006.

What is more, some of the MIS timber projects are turning what were once proven, highly profitable broad acre cropping and grazing land into considerably lower valued timber plantations.

The market distortions extend to the water market.

First, MIS timber plantations reduce the supply of water in various catchments. Intensive plantings of young trees, which absorb a considerable amount of water in their growth stage over 10-15 years, dry up surface flows and reduce ground water flows. The use of this water is at zero cost to the MIS, but it has a negative cost in reducing the water flows down streams and rivers and into reservoirs.

Second, because many MIS are flush with money from overcharging investors, they are in a position to buy up large amounts of water entitlements at above what would be normal market prices. This has the effect of raising the price of irrigation water across the system.

Third, MIS, with their financial ability to buy into the water market, are the major source of water trade among catchments. In 2005, MIS were responsible for 85 per cent of the secure water traded out of Victoria's largest water authority, Goulburn Murray Water. In 2006, water brokers have estimated that 75 per cent of Goulburn-Murray water and up to 100 per cent of Lower Murray water sold out of the catchments have been traded to just three MIS –

Timbercorp, SAI Teys McMahon and Macquarie Agribusiness.[4]

MIS lead to oversupply in various industries. This means that today's high value products can quickly become tomorrow's low value products. This makes economic modelling on water trade between catchments, from "low value" to "high value" agriculture, unreliable for policy decision-making. Arguably, economic modelling is only measuring the up side of a boom-and-bust cycle in industries facing serious over production due to large tax breaks.

Fourth, at the end of these schemes, the manager operators are left with huge water banks and sizable land banks. That water can be used for a variety of purposes, including withholding supply and forcing up water prices in drought times.

The genesis of these MIS were Federal Government investment concessions aimed at boosting investment in tree plantations, so as to cut Australia's $2 billion deficit on timber imports. Therefore, the schemes should now be limited to the timber industry. Environmental planning limitations should restrict plantations to high rainfall areas, so they do not adversely affect water use by other farmers.

Finally, it is ironical that the Federal government insists that it will not subsidise agriculture, arguing, "Why should taxpayers subsidise farmers?" Yet it is prepared to have taxpayers subsidise wealthy city investors in MIS that seriously distort agricultural markets and drive genuine farmers out of business.

Recommendation: The Federal Government must restrict managed investment schemes to timber plantations in high rainfall regions above 640 millimetres annually, and apply environmental guidelines to ensure that new plantations do not adversely reduce water flows in catchments, as well as the supply of water to other forms of agricultural and other water users.

Endnotes

1 Stephens, Mike, "Schemes are a big MIS-take", *Weekly Times,* March 29, 2006.

2 Hooper, Narelle; Anderson, Fleur; "The tax schemes that ate Australia, The Weekend *Australian Financial Review,* July 1-2, 2006.

3 Stephens, Mike, "Schemes are a big MIS-take", *Weekly Times,* March 29, 2006.

4 "Water wheels and deals leave anxious farmers dry", Carmel Egan, *The Age,* September 17, 2006.

7

Exposing the Science of Environmental Flows

The fourth pillar of the National Water Initiative (NWI) is the belief that new "environmental markets" had to be created to improve the health of degraded river systems by guaranteeing substantial flows down those rivers. Most controversial of these proposals is The Living Murray initiative, to which the federal and state governments have committed an annual environmental flow of 500 gigalitres (or 500,000 megalitres) for the river. The original proposal advocated taking up to 1,500 gigalitres for environmental flows.

This policy is controversial because there are serious questions about the science behind the proposal for environmental flows, which eventually must be at some expense to farmers. The 500 gigalitre environmental flow is equal to about 5 per cent of the 11,431 gigalitres used for irrigation in the Murray Darling Basin.[1]

Any moves towards environmental flows should be considered when and only when the proper science on river and riparian zone health has been completed. The science in favour of environmental flows under The Living Murray proposal was scrutinised closely in 2004 by the House of

Representatives Standing Committee on Agriculture, Fisheries and Forestry.

All political parties were represented on the Committee. Its interim report[2] was approved by an 11:1 majority and contained an urgent call to the Federal Government to have the Murray Darling Basin Ministerial Council, under COAG, postpone plans to commit an additional 500 gigalitres of environmental flows to the Murray River until:

- a comprehensive program of data collection and monitoring by independent scientists is completed;

- other alternatives to river management strategies, rather than just river flows, are considered and reported upon more thoroughly; and

- a full and comprehensive audit – focused specifically on the Murray-Darling Basin's water resources – including all new data, is conducted.

In order to achieve these objectives, the committee recommended that sufficient funds be diverted from the $500 million environmental funds allocated to improving the health of the Murray River by COAG.[3]

Poor science

Bill Hetherington, chairman of Murray Irrigation Ltd., claimed there was a serious absence of hard data about the health of the Murray. He said that the findings from the five scientists Murray Irrigation had employed showed

> "that salinity at Morgan has actually improved by 100 per cent in the past 20 years. There has been no change in turbidity, phosphorous and nitrate levels since they were collated in 1978. As well, the Murray cod are more plentiful than ever and carp numbers have diminished considerably. The water quality to our irrigators ... is a top world standard. So ... what is wrong?"[4]

Dr Jennifer Marohasy, of the Institute of Public Affairs Environment Unit, said there was a serious absence of evidence to support popular urban myths about a serious decline in the river's health. Adding to Mr Hetherington's evidence, she said that:

- irrigators take about 34 per cent of total inflow into the river system on average;
- approximately 41 per cent of inflows actually flow to the sea in an average year – quite a bit more than is represented by scientists or the media;
- water tables had dropped significantly in substantial areas of the basin, and in the last 20 years the amount of land impacted by shallow water tables had dropped from 127,000 ha to 14,000 ha;
- a number of studies of macroinvertebrate populations had demonstrated healthy and diverse populations, which was at odds with the conclusions drawn in the *Snapshot of the Murray-Darling Basin River Condition*, based on computer modelling; and
- the available evidence does not substantiate decline of red gum populations along the Murray.

Dr Marohasy postulated that the cause of misconceptions about the health of the Murray River was the tendency of scientific reports to ignore the occurrence of natural extreme variations in river conditions. She said:

> "We have not really thought through the implications of 'natural' as opposed to 'healthy' in the context of an old river that runs through a semi-arid environment. In such an environment, during the inevitable frequent droughts, 'natural' logically equals dead fish and stressed red gums as surface water recedes and groundwater levels drop.
>
> "Our scientists are currently compiling environmental indicators of river health all-the-while making their comparisons with hypothetical pristine environments where 'pristine' falsely equals 'well watered'. If, instead, we set our management goal as improving trends based on current

A Model for Resolving Environmental River Issues

The Barmah-Millewa Forest process provides a ground-breaking model for solving a range of environmental conflicts. This area is an important ecological wetland and forest on the Murray River between Cobram and Echuca.

Over a decade ago, a meeting between the Murray-Darling Basin Commission (MDBC) and local communities brought together a balance of farm irrigators, representatives of the departments of water resources, environmental groups, local councils, river recreational-users' groups, the aboriginal community and local councils. Special recognition was given to the groups that financially supported the management of the system.

The purpose was to find ways to flood the area periodically to regenerate the forest and wildlife, and to improve river health.

When it started, nobody believed the process could work, because of the antipathy among the many diverse groups forming the committee. Yet, after a difficult three-year process, a watering regime was agreed to.

An agreement was struck to periodically add 50 gigalitres of water to the flood flows through the region when the Ovens River flood came through this reach of the Murray. It would require the building of several regulators so that different areas of the forest would be flooded in different years. River gums need periodic, but not annual, floods for regeneration.

This process could be described as a "community cooperative sustainable self-management system." It was:

- a "cooperative" process rooted in the community, so all sides received a full hearing of their concerns.
- based on community-agreed science, not partisan science of a section of the community, providing a common understanding of the needs of this wetlands system.

- "sustainable" in that it was designed to preserve the resources of the region and to manage their use sustainably.
- "self-managed" not government-dictated, in that the management of the process was done by the whole community.

The community cooperative sustainable self-management system process worked because it led to even the most diametrically opposed groups coming to appreciate each other's concerns and the scientifically determined environmental needs of this wetland.

Most importantly, once the committee process started, no one group could make outlandish, overstated claims for their pet issues, without losing credibility in the committee process. Conservationists could not overstate their case. Farmer irrigators could not overstate their needs compared to that of the wetlands system.

Contrast this co-operative approach to resolving an environmental dispute with the cumbersome way government departments and their statutory authorities try to settle such disputes.

That process ends up with aggrieved parties, themselves lacking comprehensive knowledge of the real issues, venting their anger at opposing stakeholders and governments. The result is that governments can then be left captive to one interest group, regardless of whether that group has sound science to solve the problems.

The cooperative process should be applied to Murray River issues, where the Federal Government has decided to restore to pristine health six icon sites. Environmental issues vary from site to site, and require a large amount of detailed science and comprehensive local historical knowledge.

A "one size fits all" environmental policy won't work. The government's aim should be to get the process right, to set up individual icon site committees with all stakeholders involved, modelled on the Barmah-Millewa forest process.

conditions (that is, a healthy working river), then the issue of trying to estimate the natural or pristine environment becomes redundant ..."[5]

The committee noted that when the Murray Darling Basin Commission capped water diversions from the river system in 1994-95, an opportunity was missed to put in place research programs to capture data on improvements in river health. A decade of valuable data to guide future management of the river was not collected.

Dr Lee Benson, of Ecology Management, questioned the integrity of The Living Murray process, pointing out that in the absence of such data, the Murray-Darling Basin Commission has reverted to relying on expert panels that can do no more than guess at what makes for better river health.[6]

Issues in river health

Expert panels are no substitute for basic data. Dr Benson said that stakeholders along the Murray have a close understanding of the environmental issues and must be involved more in the practical issues of managing the river's health.

Improving the environmental health of the Murray-Darling Basin's rivers requires a complex response, involving analysis of the costs and benefits of possible variations to current management practices relating to 22 issues of river health, including:

- instream habitat: the logs, water plants, water turbidity and temperature that affect river life;
- riparian zone health, relating to stream bank stability, land and vegetation adjoining the river like wetlands and billabongs, and flood effects on the regeneration of the flora and fauna;
- instream structures: the siting and management of locks, dams and weirs that affect river flow, irrigation use and riparian zone flooding;
- seasonality of flows: the natural regeneration cycle is in

July-September (coinciding with the periodic, traditional snow melt leading to river flooding), whereas main flow timing is November-February when farmers irrigate;
- salinity management in each catchment area;
- control of pest species;
- losses of water in the distribution channels and impoundments;
- the volume of water flows down the rivers.

Conclusion

The House of Representatives Agriculture Committee concluded:

> "The level of disagreement between scientists is itself cause for concern. Of greater concern is the weight of evidence against the scientific reports.

> "The Committee asks 'would scientists promoting new treatments or pharmaceuticals to address the health problems of human beings be so cavalier in terms of paucity of data and testing as appears to be the case with the decision making process associated with the health of the Murray-Darling Basin?'"[7]

Clearly, the focus on river health has been very narrowly restricted to environmental flows when the scientific evidence has failed to show the need for such flows. Buying water from farmers for these flows will cut the amount of water available for irrigation and force up the price of water in the basin.

The policy of a 500 gigalitre annual environmental flow should be put on hold until comprehensive science studies on river health are undertaken.

Recommendation: It is imperative that governments implement the findings of the House of Representatives Standing Committee on Agriculture, Fisheries and Forestry *Inquiry into future water supplies for*

Australia's rural industries and communities – Interim Report:

Recommendation 1: In light of the Committee's severe reservations about the science, the Committee recommends that the Australian Government urge the Murray-Darling Basin Ministerial Council to postpone plans to commit an additional 500 gigalitres in increased river flows to the River Murray until:

• a comprehensive program of data collection and monitoring by independent scientists is completed;

• non-flow alternatives for environmental management are considered and reported upon more thoroughly; and

• a full and comprehensive audit focused specifically on the Murray-Darling Basin's water resources, including all new data, is conducted.

Recommendation 2: The Committee recommends that the Australian Government ask the Murray-Darling Basin Ministerial Council to allocate sufficient funds out of the $500 million allocated to the River Murray by COAG to the abovementioned tasks, prior to proceeding with the proposal to obtain increased river flows.

Recommendation: Where thorough, community agreed science shows that there is a need for environmental flows in rivers, then governments must create a separate environmental water market. Water and land set aside for environmental purposes are public goods. Water for these flows is to come from water savings from infrastructure improvements and from tapping new water supplies, not from purchasing irrigation water from farmers. The cost of environmental flows should be borne by the whole population collectively, as they are the environmental custodians of our river environments.

Endnotes

1 *The Living Murray: A discussion paper on restoring the health of the River Murray,* Murray-Darling Basin Ministerial Council, July 2002, p. 1.

2 *Inquiry into future water supplies for Australia's rural industries and communities - Interim Report,* the House of Representatives Standing Committee on Agriculture, Fisheries and Forestry, March 2004.

3 Ibid.., p. IX

4 Ibid.., p. 3.

5 Ibid.., p. 5.

6 Ibid.., p. 7.

7 Ibid.., p. 13.

8

Water Trading and the Fate of Agriculture

The four pillars of the National Water Initiative's water trading policy have taken on a life of their own. Water trade has become a mantra. It is regarded by the federal and state governments, the Productivity Commission and other think tanks as the solution to urban, industrial and environmental water shortages. It is regarded as a means to boost agricultural output. The policy is being rapidly implemented. Yet the NWI is fatally flawed.

First, it is attempting to use water trade in open markets to control the *primary allocation* of water between agricultural, urban and other uses. In reality, this can be done only by governments. Attempting to use markets for this purpose will always see urban areas win out over farmers, forcing up the price of irrigation water.

Second, contrary to the assumption by all the advocates of water trade, water is not a private good. It is a mixed good with many public good characteristics. The failure to recognise this has been a major reason for the flawed NWI policy debate.

Third, it is assumed that water trade is needed to help boost "high value" agricultural production. This is unnecessary. Farmers already respond to market forces and change agriculture to high value products; they don't need water trade to boost agricultural production.

Further, economic modelling, which claims to show a significant increase in farm output from water trading, appears to be fundamentally flawed. Most of the secure water being traded is going to managed investment schemes (MIS). These investments are not responding to market forces but to large government-sponsored tax breaks. This is leading to over investment, turning "high value" agricultural industries into "low value" industries. This produces a snow balling effect, whereby many farmers are forced to leave the industry and their irrigation channels become uneconomic causing the stranding of water infrastructure assets and the assets of other farmers. In turn, as the water supply is cut off, the remaining farmers also go out of business.

Fourth, the science to justify environmental flows down various rivers, such as the Murray, is yet to be done. The only serious scrutiny of environmental flows has been done by the federal House of Representatives Standing Committee on Agriculture, which expressed astonishment at the poor science behind the proposals for large environmental flows down the Murray River. Its recommendations have been ignored by federal and state governments.

There is no need for trading secure water out of irrigation areas. Trade from low to high value agriculture is a mistaken concept, while governments are yet to produce the scientific justification for environmental flows. The loss of water to MIS and environmental flows will force up the cost of irrigation water, contrary to the intentions of the NWI.

The water trading debate has been dominated by people who show a serious ignorance of the issues involved. Worse still, advice to governments appears to be moulded to what governments want to hear, rather than what they should hear. Recently, the federal Auditor-General, Tony Harris, expressed concern at the $300 million the Federal

Government is spending annually on outside consultants. Commenting on this, the *Australian Financial Review* said:

> "... amid concerns about the growing politicisation of public sector advice to government, there are fears that more external consultants are being tempted to provide advice that their clients want, rather than what they might need to be told."[1]

This appears to be the exact problem with advice being given to governments on NWI water trading policy.

The fate of farming

Recent data on the state of Australia's farmers reveals their precarious existence. The Australian Bureau of Agricultural and Resource Economics (ABARE) recently published trend data showing farm returns have been declining over the long term and, since about 1990, returns have been below costs most years. The trend line anticipates a further decline of farm incomes.[2]

At the same time, the Reserve Bank has revealed that rural debt has almost doubled from $26.4 billion in 1999 to $43.3 bn in 2005.[3]

Water trade will be the last straw for many farmers.

Farmers are left asking the question: does Australia want to maintain a large agriculture sector, which along with the agricultural dependent input and output industries constitute 12.2 per cent of the Australian economy and a significant proportion of our export dollars?[4]

Australian irrigation agriculture was purpose built to make Australia self-sufficient in food production, to provide low cost food to Australian households, and to create an export market, which was needed to pay for imported capital and manufactured goods not available domestically. Governments built huge irrigation systems, investing in large infrastructure that the private sector would regard as too big an investment with returns too long term for them to handle. These irrigation systems were to provide secure, reliable water at delivery cost, allowing farmers to invest heavily in high-technology agriculture.

Low cost agriculture has the objective of providing low cost quality foods to Australian consumers, while giving farmers a competitive edge on world markets.

Low cost water is part of Australian farmers' competitive advantage, on both the domestic and export market. In both markets they are competing with heavily subsidised and dumped products from other first world agricultural producers. The Organisation for Economic Co-operation and Development (OECD) tracks farm subsidies across the developed world. On average, farmers in the developed world receive 31 per cent of their gross farm income from various producer supports. In contrast, Australian farmers receive just 4 per cent of their gross income in the form of subsidies.[5] This means Australia's competitors have a 27 per cent head start when selling onto world markets, or when selling into the Australian market.

Governments seem to have lost sight of what farmers provide to the nation. The OECD compares the prices farmers in different countries receive for their products and the price consumers pay for food in different countries.[6] It shows that among the developed nations:

- Australian farmers receive the lowest farm-gate price in the developed world, and on average that price is equal to the corrupt world price. On average, farmers in OECD countries receive a price 32 per cent higher than what Australian farmers receive; and

- consequently, Australian consumers benefit from having the lowest priced food in the developed world. On average, consumers in other OECD nations pay 37 per cent more for their food than Australians.

Forced to compete with subsidised foreign products, it should be stated again; the Productivity Commission has shown that farmers make up the third highest productivity sector in the Australian economy, after the communications and wholesale sectors.[7] Reliable, low-cost water is critical to productivity.

Conclusion

A large part of Australian agriculture is facing serious financial difficulty. Undoubtedly, the drought will finish off many farmers. The NWI water trading policies are set to finish off many more. It will tear up an intricately devised water entitlements system, comparable to the system of common law.

In the face of this impending crisis, the silence of many farming organizations has been deafening. Some have been compliant supporters of COAG's NWI policies.

In the final analysis, if the current NWI policies succeed in being implemented, it won't be solely the result of the appalling ignorance of government policy makers; it will be because farmers' representative organizations failed to stand up, debate and fight for their members' water rights.

Unless farmers make their voices heard and reverse these policies now, water trading will lead to city based investors capturing control of sizable amounts of irrigation water and to the diversion of even more farm water to cities and for environmental flows down rivers. In turn, this will force up the price of water, drive more farmers off the land and cause food prices to rise for consumers in the cities.

It is time to make politicians and policy advisors confront the disaster their policies are unleashing on Australian agriculture.

Endnotes

1 Hughes, Duncan, "Call to cut spending on consultants", *Australian Financial Review,* October 2, 2006.

2 Hunt, Peter, "Incomes on downhill slide, *The Weekly Times,* August 2, 2006.

3 Le Grand, Danielle, "Farming debt binge", *The Weekly Times,* May 31, 2006.

4 *Australia's Farm Dependent Economy: Analysis of the Role of Agriculture in the Australian Economy,* research paper by the Australian Farm Institute, 2005.

5 *Agricultural Policies in OECD Countries Monitoring and Evaluation 2003,* Figure 2.2, p. 29.

6 Compiled from *Agricultural Policies in OECD Countries Monitoring and Evaluation 2003,* Table Annex 2, PSE by Country pp. 44-45 and Annex Table 3, PSE by Commodity, pp 46-47. See unpublished research paper, "Regions between theory and reality: Agricultural policy and its impacts in Australia", by Ben Rees and Mark McGovern, School of International Business, Queensland University of Technology; 1994.

7 Banks, Gary, Productivity Commission Chairman, *The drivers of Australia's productivity surge.* Paper Presented at Outlook 2002, hosted by the Department of Industry, Tourism and Resources and the Australian Bureau of Agriculture and Resource Economics, National Convention Centre, Canberra, 7 March, Figure 4, p. 6.